Contents

2 Introduction

4 Make a tent

6 Bridge that gap

8 Spice it up

10 Rolling along

12 Out of the window

14 Do all clocks tick?

16 Racing cars

18 Spot the difference

20 Eggs-change

22 Butterfly cycle

24 Leaf patterns

26 Colour splodges

28 Choose your tools

30 Turning tops

32 Skills development chart

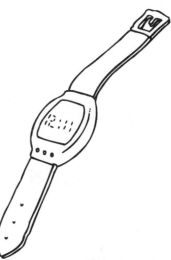

Introduction

Developing science skills

Early skills that help children to focus their natural interest and curiosity in the world around them involve investigating, observing, exploring, problem solving, predicting, discussing their ideas and explaining why things happen. This book contains a range of practical activities designed to encourage the

development of these skills. Each activity will provide the children in your group with opportunities to explore and investigate, to talk about their observations and to begin to develop scientific thinking as they come to their own conclusions. The practical tasks are fun to carry out and are designed to foster further investigation. Some of the activities can be used with small groups of children, some slightly larger and some with the whole group. This variation of group sizes provides the children with different ways of working together and enables them to share their growing knowledge in order to arrive at some scientific conclusions.

Early Learning Goals

The science skills included in this book form part of the Knowledge and understanding of the world area of the Early Learning Goals identified by the Qualifications and Curriculum Authority (QCA). The ideas suggested in this book can be applied equally well to the documents on pre-school education published for Scotland, Wales and Northern Ireland. The activities are divided into four sections that mirror the requirements for this area, namely:
• to investigate objects and materials using all of the senses as appropriate
• to find out and identify some of the features of living things, objects and events they observe
• to look closely at similarities, differences, patterns and change
• to ask questions about why things happen and how things work.

A lot of early science skills also involve the other Areas of Learning – for example, the children will also be developing mathematical skills when measuring, ordering and discussing shapes and sizes. All science skills promote language development and several skills also include aspects of early reading and writing. Many of the activities foster the children's creativity by allowing them to respond to tasks in their own way. The science activities also give the children plenty of opportunities to use small tools and to handle small objects, thus promoting good hand control, which fosters the development of fine motor skills.

Baseline Assessment

Each activity includes suggestions for independent recording and ways in which individual assessment can be carried out as the children complete each task. These can form the basis of future records of achievement, baseline assessment and for individual profiles. This is in addition to the completed photocopiable sheet that accompanies each practical activity. Where a photocopiable sheet is intended to be used by an individual, it is referred to as an 'Individual recording'. If it is to be used by a group of children, then it is referred to as an 'Individual task'.

How to use this book

Part of a series that provides new ideas for developing early learning skills, this book is designed to help all those working in a pre-school setting to provide some of the experiences needed to enable children to work through the Stepping Stones of the Foundation Stage relating to the Knowledge and understanding of the world area of the Early Learning Goals.

All the activities make use of resources that are normally readily available in pre-school settings or can be easily obtained. Preparation time has been kept to a minimum and the assessment tasks are designed to be manageable within the normal working day. All of the activities are designed to be used according to the children's individual levels of development. At the end of each activity, you will find ideas to support those who are younger or less able and suggestions to extend the activity for use with older or more able children.

The last page of this book contains a photocopiable 'Skills development chart'. As each skill is introduced, the adult and child can colour in the relevant part of the chart together. This helps to foster a sense of ongoing achievement, and the chart can be used as part of individual children's record folders.

Home links

Each activity contains suggestions to continue or extend the learning at home. This will enable parents and carers to become more closely involved in the tasks their children do at your setting and will provide them with a good insight into their children's learning. It will also reinforce the skills learned at your setting by developing them in the home environment.

Try to provide plenty of opportunities for the carers to become involved. Let the children take home their work to share with their families, and inform parents and carers of your forthcoming activities so that they can contribute resources, or even come in to your setting to help out with the preparation or the tasks. Collaboration between home and pre-school provides the best possible foundation for the children's learning.

Make a tent

Learning objective
To investigate the waterproofing properties of different materials.

Group size
Up to six children.

What you need
A copy of the photocopiable sheet for each child; scissors; glue; a selection of waterproof and non-waterproof materials such as cotton, paper, corrugated card and plastic (all approximately 10cm x 10cm); small tray of water; string or washing line; pegs; paper clips; jug; newspaper; large water tray.

Preparation
Use the tepee template to cut several discs of each of the different materials (at least one piece per child). Cover the table with newspaper. Suspend the string or washing line at the children's height.

What to do
Sit the group around a table with the selection of fabrics and a tray of water in the middle. Let the children handle the fabrics freely and talk about the different textures. Encourage them to use descriptive language such as 'shiny', 'slippery' and 'furry'. Discuss which fabric might be best for making a tent. Which might be most effective at keeping out the rain? Why?

Select some of the pieces of fabric and dip them into the water. Let the children peg the wet pieces on the line. What has happened to each one? Was this what the children expected? Explain that each child is going to make a tent from a different fabric and then all the tents will be tested for waterproofness. Invite each child to choose one of the pre-cut discs of fabric to use for their tent.

Individual recording
Show the children how to cut out the tepee template from the photocopiable sheet and glue their chosen piece of material onto it. Help them to cut along the line, fold over and secure with a paper clip so that each child has a material-covered tent. When all the tents are complete, gather around the water tray. Show the children how to pour a small amount of water from the jug over their tents. What happens to the water and to the fabric? Which child's tent is the most waterproof? Was this what they expected?

Encourage the children to complete their photocopiable sheets by gluing samples of the materials onto them, showing three of those that were tested and the most waterproof out of those three.

Support
Cut out the tepee template for younger children and provide plenty of assistance with the gluing and pouring of water.

Extension
Let older children make additional tepees and set up a model village for imaginative play.

Assessment
Check whether the children have an understanding of which materials are waterproof and if they are aware of why that is important when building a home. Can they say why the shape of the tepee is also important?

Home links
Ask parents and carers to help their children to make a collection of waterproof things found in the home.

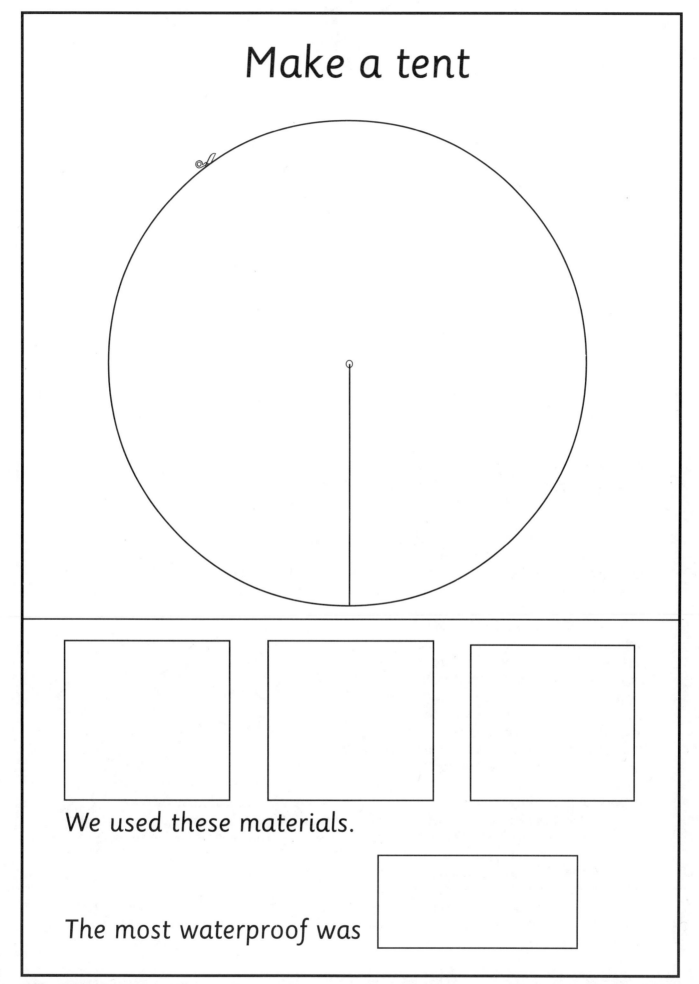

Bridge that gap

Learning objectives
To investigate different materials; to think about why things happen and how things work.

Group size
Two to four children at a time.

What you need
A copy of the photocopiable sheet for each child; wooden bricks; paper; thin card; thick card; toy cars; train track; empty floor space.

Preparation
Cut the paper and card into lengths measuring approximately 8cm x 25cm. Arrange the train track in a circle on the floor.

What to do
Sit on the floor with the children and show them the wooden bricks, train track, paper, card and toy cars. Explain that you would like their help to get the cars across the train track, using the resources available. Play with the children as they explore the different resources. Ask questions to direct their play, such as, 'How do cars usually get across train tracks?', 'What could be used to make a bridge?' and so on.

Encourage the children to select wooden bricks to make two supports of equal height for the bridge. Once the supports are in place, suggest using the paper and card to bridge the gap. Investigate each one in turn. Does the paper support the weight of the cars? Why not? What about the thin card? Lastly, try the thick card. Which bridge was the strongest?

Individual recording
Invite the children to look at each picture in turn on their photocopiable sheet and to decide what each bridge is made from, drawing on the conclusions that they came to during their play. Help them to write the correct material name to finish each sentence, and then to complete the sentence at the bottom of the sheet to indicate the strongest bridge.

Support
Help younger children to explain their ideas and suggest new vocabulary if needed.

Extension
Encourage older children to try driving a heavier vehicle such as a toy lorry across the different bridges. What do they think will happen? What actually happens? Can the children suggest a stronger material to use that would support the weight of the lorry?

Assessment
Note how the children approach the task. Are they keen to find a solution? Do they show a good awareness of the purpose of bridges and suitable materials?

Home links
Ask parents and carers to help their children notice bridges that they pass under or over when they are out in the local environment. Invite them to encourage their children to notice the different shapes of the bridges and to make drawings of these when they return home.

Bridge that gap

This bridge is made from _____

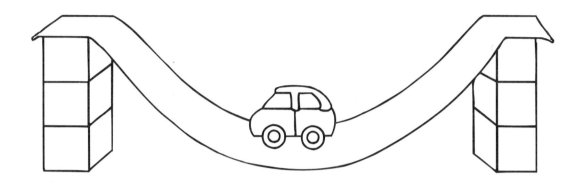

This bridge is made from _____

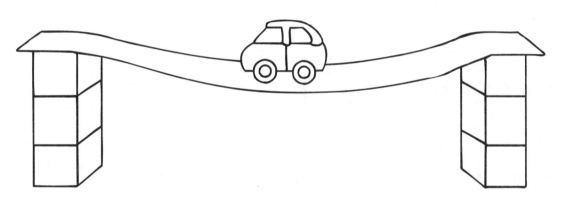

This bridge is made from _____

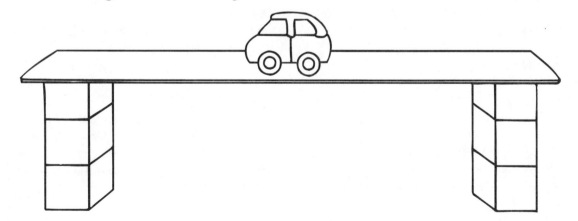

The strongest bridge is the one made from

_____

Spice it up

Learning objective
To investigate objects using all the senses as appropriate

Group size
Six to eight children.

What you need
An enlarged copy of the photocopiable sheet on A3 card; recipe ingredients (ideally with the spices in jars or tins with labels); frying pan; small knives (to be used under adult supervision); cutting boards; aprons; access to a cooker; serving plates.

Preparation
Ensure that there are cooking and washing facilities to hand.

What to do
Start by asking the children to wash their hands and put on their aprons. Talk briefly about hygiene and safety when cooking. Sit together around a table and explain the task. Introduce the spices and pass each around the group so that the children can smell them in turn. Encourage the children to describe the different smells and talk about how the spices are different from one another. Name the spices, pointing out the name labels on the spice jars or tins. Then place the vegetables on the table and talk about the features that they share and what makes them vegetables. Encourage the children to name them with you.

Give each child a vegetable and provide hand-over-hand support to help them cut it up on their board. How does each vegetable smell? Let the children taste some that can safely be eaten raw, such as carrot and onion. (NB: Check for allergies and dietary requirements). How do they taste?

Individual task
Show the children the recipe on the A3 card. Follow it together, inviting everyone to help add the ingredients. Once all the ingredients are in the pan, ensure that an adult continues the cooking process and that the children are kept at a safe distance. Encourage them to describe the smell of the curry as it cooks. Once it is cooked, serve small amounts to each child with a little naan bread. What does the curry taste like? Can they recognize the vegetables now?

Support
Provide pre-cut vegetables for the children to investigate.

Extension
Set up an interest table with books, pictures and a variety of vegetables and spices for the children to look at in their own time.

Assessment
Note whether the children are able to name any of the vegetables and spices. Are they willing to try a variety of tastes? Can they describe the different smells and tastes that they experience while carrying out the activity?

Home links
Ask parents and carers to discuss the smells and tastes of foods eaten at home and to give their children an unusual food item to bring in for the interest table.

Spice it up

1 tbsp oil
1 onion
½ tsp turmeric
1 tsp ground coriander
½ pint water

cauliflower pieces
2 carrots
1 potato
1 courgette
½ lb spinach
salt and pepper

What to do

■ Chop the onion into small pieces. Heat the oil in a large pan. Cook the onion until soft (not brown).

■ Add turmeric and coriander. Cook for 1 minute, stirring continuously.

■ Add the water, salt and pepper. Stir slowly to make a smooth gravy texture. Simmer for 1 minute.

■ Cut up the vegetables as finely as possible. Add to the sauce and cook until tender.

■ Add the washed spinach leaves and stir. Cook for 5 minutes.

■ Serve!

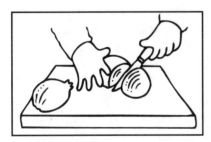  
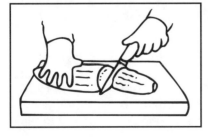  

Rolling along

Learning objective
To investigate and identify the movement of different-shaped objects.

Group size
Four children.

What you need
The photocopiable sheet; collection of toys such as toy vehicles, construction bricks, balls, play people and so on (ensure that your collection includes some toys that roll and some that do not roll); colouring materials.

Preparation
Make a copy of the photocopiable sheet for each child.

What to do
Put the collection of objects on a table and invite each child to choose one to examine. Allow plenty of time for the whole group to look at all the objects and talk about them together. Invite the children to try in turn to roll their objects across the table. Work together to place the objects into two sets: those that roll and those that do not. Ask questions as you sort the objects, such as, 'Why does the marble roll?', 'What does the toy car have that makes it roll?', 'Why does the building block not roll?' and so on. When all the objects have been sorted, look at each group. Can you draw any conclusions about the group of objects that roll? What about those that do not roll? Notice that the rounded objects roll more easily. Why might this be?

Sing 'The Wheels on the Bus' and talk about what the words 'going round' mean. Encourage the children to make the rolling movements with their hands as they sing.

Individual recording
Invite each child to identify the items on their photocopiable sheet that roll and to use a pencil to draw a circle around each one. Make the resources available so that the children can test their answers. Were they correct? When they are happy with their answers, invite them to draw over their pencil lines using a colourful felt-tipped pen. Finish by colouring in the pictures.

Support
Limit the objects available to spheres, cylinders and cuboids. Help the children to realize that shapes with corners are harder to roll than the shapes with curved diameters.

Extension
Provide materials such as fabric, Plasticine and cardboard. Can the children mould or bend these into shapes that roll?

Assessment
Note whether the children can correctly identify the objects that roll. Can they offer appropriate suggestions about the reasons why some objects roll and some do not? Can they circle the correct objects on the photocopiable sheet?

Home links
Ask parents and carers to help their children to find examples of one thing that rolls and one thing that does not to bring in for a table display.

Rolling along

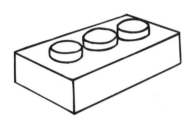 

Out of the window

Learning objective
To observe and identify features in their environment.

Group size
Groups of four or five children at a time.

What you need
A copy of the photocopiable sheet for each child; paper; pencils; clipboards.

Preparation
Choose a window at your setting with an interesting view. This might be of a bird-table, garden area or busy street.

What to do
Gather the children together by the window area. Spend some time looking at the view. Ask the group questions such as, 'What can you see?', 'Are there any trees?', 'Can you see any traffic?', 'Can you see any people?', 'What can you see in the sky?' and so on. Make a list of the children's suggestions. Give each child a clipboard and a piece of paper and encourage them to draw simple pictures of some of the things that they can see out of the window.

Independent recording
Give each child a copy of the photocopiable sheet. Invite them to look carefully at the key at the bottom of the page. Using their drawings, and the list of objects that you wrote down, can they identify any of the things that they saw out of the window? As they identify items, invite them to colour in the individual pictures. Next, encourage them to look at the main picture and to find the items from the key, then to colour these so that they match the individual pictures.

Support
Instead of giving the children plain paper on which to record their observations, provide simple pictures of things that they are likely to see. Support them as they look for these items outside, and help them to tick off each item as they identify it.

Extension
Encourage each child to add other features to the scene on their photocopiable sheet. Perhaps it was raining when you looked out of the window. If so, invite them to add rain clouds, puddles and umbrellas!

Assessment
Note whether the children are able to identify and describe the objects that they saw out of the window. Can they recall the same objects when looking at the key and the illustration on the photocopiable sheet?

Home links
Provide a sheet with boxes for the children to fill in their own key and scene. Invite them to complete the sheets at home using a view from their choice of window. Encourage them to bring them back in to your setting to discuss.

Out of the window

Key

Do all clocks tick?

Learning objectives
To find out about and identify some features of objects; to ask questions about how things work.

Group size
Six children.

What you need
A copy of the photocopiable sheet for each child; a collection of digital and analogue clocks (for example, a carriage clock, pendulum clock, alarm clock, digital stop watch, radio alarm clock and digital watch); the story *Clocks and More Clocks* by Pat Hutchins (Bodley Head); screwdriver; magnifying glasses.

What to do
Sit together in a circle with the collection of clocks in the centre within easy reach. Encourage the group to carefully examine each one in turn. Invite the children to tell you something about each clock. Pass one of the analogue clocks around the circle and ask the children to listen carefully to it. What sound does it make? Repeat with a digital clock. Does it make the same sound?

Listen to each clock in turn. Help to develop the children's investigative skills by prompting them with questions. What is similar about all the ticking clocks? Do the digital clocks have moving hands? Help the children to notice that the clocks that tick are the ones that have moving hands. Look at an analogue clock. If possible, demonstrate how to wind the clock up. Carefully open up the back and look at the wheels, cogs and springs. Explain how the spring gradually uncoils second by second to record the passing of time, and this is what causes the tick. Let the children use the magnifying glasses to examine the moving parts closely. Finish by sharing the story *Clocks and More Clocks*.

Individual recording
Give each child a copy of the photocopiable sheet and help them to cut out the individual pictures. Encourage them to sort their pictures into two sets to show clocks that tick and clocks that do not tick. Stick them onto strips of paper to make individual zigzag books.

Support
Work with smaller groups and limit the selection of clocks that you look at.

Extension
Make models of clocks with moving hands using a photocopied card clock face and two hands attached with split-pin fasteners.

Assessment
Note whether the children can identify the difference between the digital and analogue clocks. Can they correctly sort the clock pictures on the sheet into those that tick and those that do not tick?

Home links
Ask parents and carers to help their children to find clocks at home. Are they all the same type? Do they all tick?

Do all clocks tick?

Racing cars

Learning objective
To carry out simple investigations and ask questions about events that they observe.

Group size
Two children at a time.

What you need
The photocopiable sheet; bricks; two identical toy cars; two pieces of strong card – one measuring 30cm and one measuring 1m; clear carpeted area.

Preparation
Make a copy of the photocopiable sheet for each child. Fold up the sides of the cardboard pieces so that the cars do not fall off the sides.

What to do
Take the children to the carpeted area. Set up the bricks to make two bases of the same height. Give each child a piece of card and a toy car, and ask them to place their card on their base to make a ramp. Look at the two ramps. How are they different? Notice that one is longer than the other. Can the children suggest what might happen if they roll the cars down the ramps? Which car would reach the bottom first? Why?

Test the children's suggestions by letting them hold their cars at the top of the ramps. On a given signal, both children should let go. What happens? Respond to the children's questions as they watch the cars travel down the ramps and help them to draw the conclusion that the shorter and steeper the ramp, the quicker the car will reach the bottom.

Individual recording
Using the photocopiable sheet, ask each child to decide which car will reach the bottom of the ramp first and to put a tick in the correct box. Extend by looking at the two slides. Can the children now guess which child will reach the bottom of their slide first?

Support
Place a box or plank a short distance from the bottom of the ramps. This will help the children to identify which car is fastest as one will reach the 'bumper' before the other.

Extension
Use the same length of ramps but alter the height of the brick base. Which car rolls down the ramp the fastest now?

Assessment
Can the children describe what has happened and talk about the events that they observed? Can they suggest why the length of the ramp affects the speed of the car?

Home links
Explain to parents and carers that you have been looking at ramps, and ask them to take their children to a local park to have a go on a slide. Encourage them to point out the similarity of the slide to the ramps that the children made.

Racing cars

Spot the difference

Learning objective
To look closely at similarities and differences.

Group size
Six children.

What you need
A copy of the photocopiable sheet for each child; objects in sets that are similar but not identical, such as shells, tree seeds, buttons with different colours and numbers of holes, pebbles and stones; a magnifying glass for each child; flip chart; felt-tipped pens.

Preparation
Make an A3 copy of the photocopiable sheet and cut out the two butterfly pictures.

What to do
Sit with the children on the carpet and place the two large photocopied butterfly pictures in front of them. Invite them to look carefully at the pictures and make observations. Are they the same? What differences are there? Make a list of the differences on the flip chart.

Introduce your collection of objects. Give one object to each child and invite them to look closely at it using the magnifying glass. Give them time to explore the object and then ask them to describe one of its features. When you think they have observed the objects for a sufficient length of time, give each child a second object from the same set and encourage them to look closely at it, comparing it to the first object and finding differences. Invite them to tell everyone if they can 'spot the difference'.

Individual recording
Give each child a copy of the photocopiable sheet. Encourage them to find the difference between each pair of minibeasts and to complete the pictures so that each pair matches.

Support
If the children are struggling to find differences, prompt them with questions such as, 'Are there four spots on this one?', 'What about the second one?' and so on.

Extension
Encourage the children to find differences between four similar objects rather than two. Are the features always the same?

Assessment
Check that the children are looking carefully and taking time to think about the features and the differences when using the photocopiable sheet. Check that they have successfully completed each picture to make it the same as its partner.

Home links
Ask parents and carers to encourage their children to help sort things around the home which are similar, but different, such as tins of soup, cereal packets, clothes and newspapers. Invite them to discuss the different features with their children.

Spot the difference

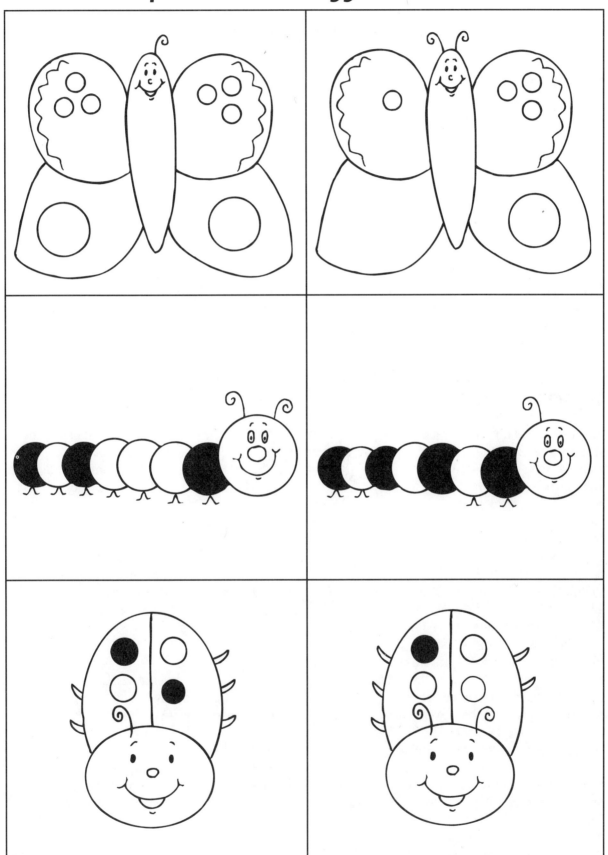

Eggs-change

Learning objectives
To investigate objects using all of their senses as appropriate; to look closely at change.

Group size
Six children

What you need
A copy of the photocopiable sheet for each child and one enlarged for group use; six eggs; bowl; wooden spoon; whisk; two small saucepans; frying pan; cooking oil; butter; aprons; cooker; paper; pencils.

Preparation
Check for allergies and dietary requirements.

What to do
Ask the children to wash their hands and put on aprons. Sit around the table with the box of eggs in the middle, then encourage the children to describe the shape and colour of the eggs.

Provide pencils and paper, and ask each child to draw a picture of an egg. Break one egg into a bowl and invite them to describe it. Encourage them to draw a picture of the inside of the egg now that it is broken. Discuss colours, shapes, smell and consistency. (NB Ensure that the children do not touch or taste the uncooked egg.) What is the difference between the outside and inside of the egg? What will happen to the egg when it is cooked?

At a safe distance, fry one of the eggs and ask the children what has changed. Encourage them to draw a picture of the fried egg. Hard-boil another egg. What do children think will happen? Is it the same as the fried egg? Crack the remaining four eggs into a bowl and let the children take turns to whisk them.

Place the enlarged photocopiable sheet where the children can see it. Prepare a frying pan with a little oil. When the oil is hot, pour in half the mixture and make an omelette, reading the instructions to the children as you follow them.

Pour the other half into a small saucepan with a knob of butter and stir with a wooden spoon to make scrambled eggs. What differences can the children see? Let them taste the different types of cooked egg.

Individual recording
Encourage the children to use their observations to colour in the pictures of the different types of cooked egg on the photocopiable sheet. Ask them which one they enjoyed eating the most.

Support
Take photographs of the eggs in their various stages if the children find drawing them too difficult.

Extension
Make a book with the pictures the children have drawn or with the photographs that you have taken. Arrange pictures in order and write captions for each to explain the process.

Assessment
Note the children's level of response to your questions. Can they say how the eggs will change when they are cooked?

Home links
Let the children take the omelette recipe home. Ask parents and carers to talk about the way that they cook eggs at home and the changes that take place. Encourage the children to find out what happens when eggs are used to make a cake.

Eggs-change

What you need
2 medium eggs
oil
salt and pepper

What to do
■ Break the eggs into a bowl. Beat them lightly. Season with salt and pepper.
■ Put a frying pan over a medium heat. When it is hot, add a little oil.
■ Pour in the eggs. Use a fork to draw the eggs into the centre of the pan, letting the liquid egg run towards the edge. Repeat until almost set.
■ Turn the omelette over and cook for 1 minute on the other side.
■ Fold over and serve.

Butterfly cycle

Learning objectives
To find out about and identify some features of living things; to investigate sequences.

Group size
Four children.

What you need
A copy of the photocopiable sheet for each child; information and story-books about butterflies, such as *The Very Hungry Caterpillar* by Eric Carle (Puffin Books) and *From Caterpillar to Butterfly* by Stewart Legg (Franklin Watts); thin A4 card; scissors; glue; hole-punch; string; colouring materials; four strips of card.

Preparation
Make four card rings by bending the strips around and stapling the ends together. These will form the supports from which the mobiles will be suspended.

What to do
Look at the books together and discuss the pictures. What changes are taking place? Discuss the order of change. Does the order repeat itself, forming a pattern of change? Read the story of *The Very Hungry Caterpillar*. Look carefully at the pictures and ask the children what they think will happen to the caterpillar.

Individual recording
Give each child a photocopiable sheet and invite them to colour in the pictures, using the books available to get an idea of the appropriate colours. Help the children to stick their sheets onto the thin A4 card then to cut out the four pictures carefully along the lines. Make holes for the children in the centre of each picture and show them how to thread the string through. Help them to attach the string to the mobile ring, tying them on in order so that they form a complete cycle. Move the mobiles around as you reinforce the cycle of egg, caterpillar, pupa, butterfly, egg and so on.

Support
Help younger children to cut out the pictures on their sheets.

Extension
Look at the life cycles of other creatures such as a frog or a chicken. Make mobiles using simple pictures or photographs.

Assessment
Note whether the children are able to put the pictures into the correct order. Do they show an awareness of change?

Home links
Ask parents and carers to talk about the changes that take place when animals and children grow. Let the children bring in pictures of themselves when they were babies to share with the group. How have they changed?

Butterfly cycle

Leaf patterns

Learning objective
To observe and identify features of natural objects.

Group size
Whole group.

What you need
A copy of the photocopiable sheet for each child; carrier bags; pictures, books and photographs of leaves; sorting trays.

Preparation
Arrange for the group to visit a nearby garden, park or wood where leaves can be found and arrange for sufficient adults to accompany the group on the walk. Label the sorting trays with pictures and words for different types of leaves – for example, spiky, curved edges, long and narrow, palmate and oval, and so on.

What to do
Share the books and pictures with the children. Discuss the different types of leaves, commenting on their shapes, colours and textures. On your chosen day, take the children on a walk and look at the trees. Talk about some trees which are evergreen and explain what that means. Talk about the other trees shedding their leaves. Let the children then hunt for leaves and collect them in their bags, guiding them to a variety of trees so that they collect different types of leaves. Remind them that they must only pick leaves that have fallen, and must not pull any from the trees.

Back indoors, sit together in a circle with the sorting trays in the middle. Ask the children to select a leaf from their bags in turn and decide together which tray it should go in. Talk about the different shapes, sizes and colours of the leaves.

Individual recording
Invite the children to look carefully at the different leaves on their photocopiable sheets. Did they find any leaves that matched the ones shown? Encourage them to sort out the real leaves and match them to the pictures. Invite them to carefully colour the pictures, using the leaves that they found for reference.

Support
Offer assistance to younger children in matching the real leaves to the pictures.

Extension
Encourage older children to use the leaves for leaf printing, leaf rubbings and to make a leaf collage.

Assessment
Check whether the children are able to see the differences between the leaves. Can they describe some of the patterns? Do they show an understanding of the difference between evergreen and deciduous trees?

Home links
Challenge the children to find different types of leaves from home or during outings with their families and to bring them to your setting. How many different varieties can you find as a group?

Leaf patterns

Colour splodges

Learning objectives
To look closely at change; to ask questions about why things happen.

Group size
Six to eight children.

What you need
A copy of the photocopiable sheet for each child; blotting paper; small tray of water; red, yellow and blue felt-tipped pens; pencils.

Preparation
Cut up the blotting paper into small pieces.

What to do
Give each child a sheet of dry blotting paper. Show them how to dampen it by carefully dipping it in the tray of water. What happens to the paper? Invite the children to talk about the changes that they observe. Introduce the term 'absorb' and discuss how the water is absorbed by the blotting paper. Let each child choose one colour of felt-tipped pen and dot a big splodge of colour onto their paper. What happens? Why do they think the colour is spreading? Let the children try the other colours in turn. What differences are there? Do any of the colours spread more than others? What happens when two colours overlap?

Ask everyone to dot a splodge of red on to their blotting paper, then to dot a splodge of yellow next to it. What happens? Repeat with splodges of blue and yellow, then splodges of red and blue. Observe as new colours are created and invite the children to name the new colours.

Individual recording
Ask the children to use their photocopiable sheets to record their findings. Which colour spreads the furthest?

Support
Use the pre-dampened sheets if the children are likely to make the paper too wet.

Extension
Experiment with blotting paper which is dry, damp, wet and soaked. Which is the best for making the colour spread? Can the children guess why?

Assessment
Can the children offer an explanation as to why the colours spread on the damp blotting paper? Do they understand that two colours mixed together make a new colour? Do they show a growing understanding of the changes?

Home links
Ask parents and carers to talk about the way that kitchen towels absorb spills and encourage them to watch the way that water is soaked up. They could also try using a small amount of cooking oil and talk about the differences.

Colour splodges

red + yellow =

blue + yellow =

red + blue =

spreads the furthest

Choose your tools

Place the A3 photocopy of the game in the centre with the tool cards next to it in a pile face down. Give each child a counter. Invite the players to take turns to throw the dice and move their counter the appropriate number of spaces around the board. If they land on a task, they have to pick up a card from the pile and describe how they would use the tool to complete the task before play can continue. If the tool on the card is not appropriate, they miss a turn. The first child to get to the end is the winner.

Individual recording
Give each child a photocopiable sheet. Encourage them to colour in the board and the tool cards, then to cut out the tool cards very carefully and to match them to the correct tasks. Let them play the game again in pairs.

Learning objective
To select tools and techniques needed for various tasks.

Group size
Groups of four working with an adult.

What you need
A copy of the photocopiable sheet for each child and one enlarged onto A3 card; scissors; glue; colouring materials; dice; counters; a collection of small tools to match those on the sheet; laminating materials.

Preparation
Colour in the enlarged board game and tool cards and laminate. Cut out the tool cards separately.

What to do
Sit the children in a small group around a table with the pile of small tools in the centre. Let the children carefully examine each one in turn. What could they use the scissors for? What jobs might they need the dustpan and brush for?

Support
Help younger children to match the tool cards to the correct tasks. Discuss why the tools are appropriate as you work together.

Extension
Let older children think of more tasks and tools to extend the game.

Assessment
Ask the children to explain how each of the tools in your collection is used and then check that they are able to match tools to the tasks in an appropriate way using the photocopiable sheet. Make a note of their responses.

Home links
Ask parents and carers to talk to their children about the tools and equipment they use in the home and how these items work.

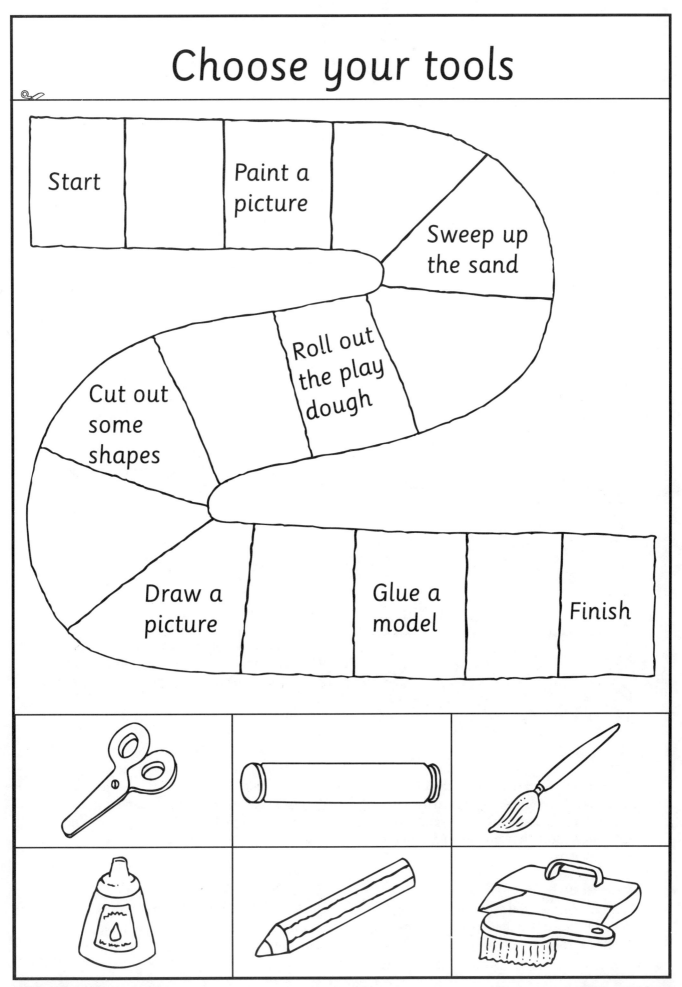

Turning tops

Learning objective
To investigate similarly shaped objects; ask questions about how things work.

Group size
Six children working with an adult.

What you need
A copy of the photocopiable sheet on card for each child and one copy for yourself; wheeled toys; clock with a second hand; hand whisk; wooden spinning top; card; glue; felt-tipped pens; scissors; spent, unbroken matchsticks (long type); a skewer or knitting needle (adult use only).

Preparation
Colour in the sections of one of the wheels on the photocopiable sheet using six different colours, then skewer a hole through its centre. Push a matchstick through the hole to make a colour spinner.

What to do
Put all the items in the centre of the table. Give each child one of the items and ask them to say something about it. Ask what the items have in common. Does each one move? If so, how? Why? Draw attention to the spinning top and the way it turns. Encourage each child to have a go at spinning it. What happens when it spins? Show the children the colour spinner that you made earlier. Does it turn in the same way? What happens to the colours when it spins?

Individual recording
Invite each child to colour in one of the colour spinners on their own sheet using six different colours. Help them to cut around the circle, then skewer a hole in the centre for them. Let them insert a matchstick through the hole and have a go at spinning it. Encourage them to describe what happens to the colours. When they have had time to observe and discuss the spinner, ask the children to carry out a similar exercise on the second spinner, but this time using just two colours. What happens now?

Support
Help the children with the cutting out if they find this difficult.

Extension
Make other tops using different designs. Colour the sections so that opposite colours match. What happens? Draw similar pictures in each space. Do the pictures appear to move when the top is spun?

Assessment
Note the children's answers to the 'how' and 'why' questions. Do they have a growing understanding of the fact that the top of the spinner needs to be spun in order to move? Do they notice that wheels, whisk and clock cogs also turn in a circular motion?

Home links
Ask parents and carers to encourage their children to find at home examples of things that turn and to select an item to bring in to your setting.

Turning tops

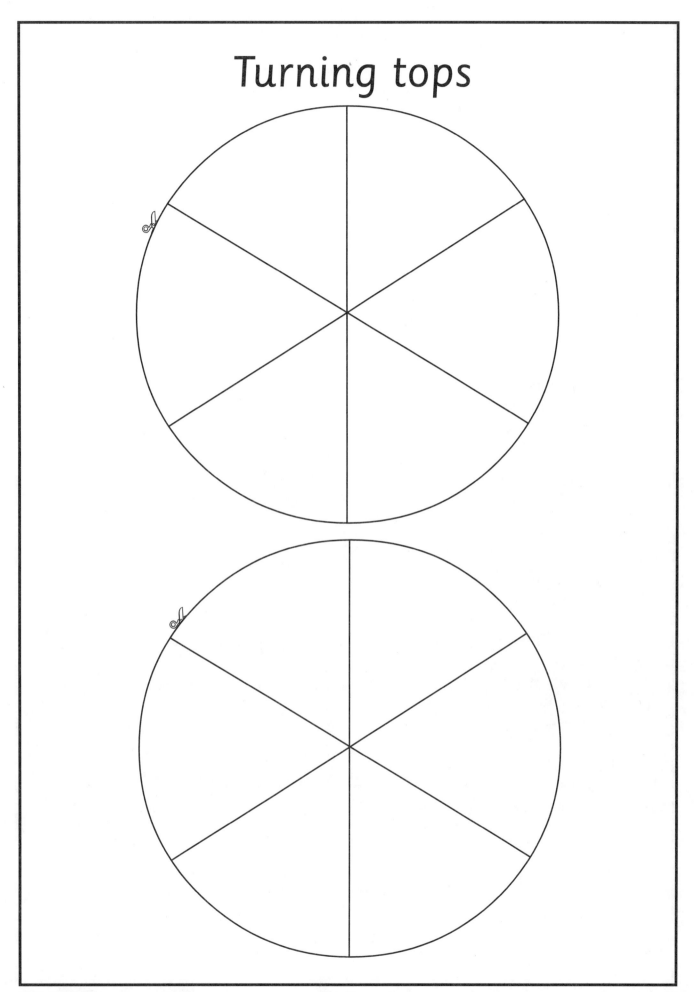

Name _____

Skills development chart

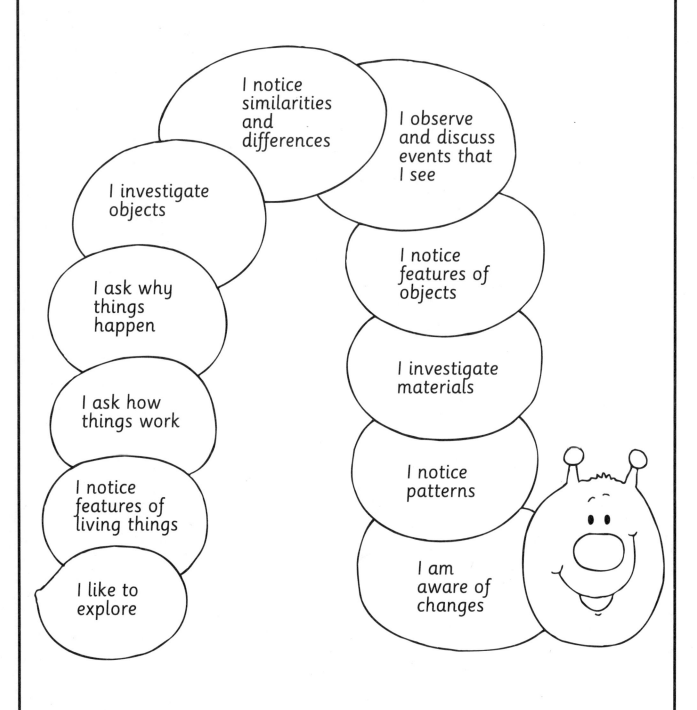